W9-BHJ-024

# INTERNSHIP & VOLUNTEER OPPORTUNITIES

## for People Who Love Nature

Greg Roza

HOWLAND MIDDLE SCHOOL

ROSEN
PUBLISHING®

New York

Published in 2013 by The Rosen Publishing Group, Inc.
29 East 21st Street, New York, NY 10010

Copyright © 2013 by The Rosen Publishing Group, Inc.

First Edition

**Library of Congress Cataloging-in-Publication Data**

Roza, Greg.
Internship & volunteer opportunities for people who love nature/Greg Roza.
    p. cm.—(A foot in the door)
Includes bibliographical references and index.
ISBN 978-1-4488-8297-7 (library binding)
1. Environmental sciences—Vocational guidance. 2. Volunteer workers in environmental protection. I. Title.
GE60.G69 2013
363.70023—dc23

2012011893

*Manufactured in the United States of America*

CPSIA Compliance Information: Batch #W13YA: For further information, contact Rosen Publishing, New York, New York, at 1-800-237-9932.

# Contents

At the Houston Zoo, an intern cuddles a newborn Chilean flamingo. This young woman's love for animals surely helped her land this internship and improves her chances of starting a career in animal care.

# Introduction

I f you keep up on current events, you've likely heard a lot about environmental problems. In a world where technology and industry are so important, our planet, its habitats, and its wildlife have had to endure great problems in the past one hundred years. It's hard to deny that human activities have taken a toll on the natural world. For example, in 2010, the *Deepwater Horizon* oil platform off the coast of Louisiana exploded, causing the largest offshore oil spill in U.S. history. The results were catastrophic to local wildlife. Luckily there are many concerned professionals hard at work to protect and restore the marine habitat.

The world needs more workers with a greater sense of environmental stewardship. It takes specially trained professionals with a true love of nature to help reduce the negative effects of human industry and ensure Earth will be able to sustain future generations.

Whether you're into gardening, caring for animals, camping, or teaching, there are numerous jobs available for people who love nature. National parks and wildlife preserves need forest rangers. Or if you're more interested in marine habitats, you could pursue a career in marine biology. Whatever your natural interests may be, there's a job for you.

However, becoming a forest ranger or marine biologist isn't as simple as applying for a job. Both positions require a college degree. Though there are nature jobs that don't require a college degree, many will still require you to have prior experience in the field. This is where internships and volunteer experiences will help you prepare.

An internship is a temporary, usually unpaid position with an organization. Interning is a typical path for college students, particularly those who are close to completing their degrees. Often, an internship results in a full-time position with the same organization after graduation.

Some internships can last for several months while others span just a few weeks. They're usually structured learning experiences that prepare you for a college program or a career. There are also numerous internships available for high school students, many of which will allow you to earn college credits.

Volunteering is a less structured, but no less fulfilling method of gaining valuable experience. This path is a great choice for young people who aren't ready for college. It's also a great way for college students to supplement their education outside of class.

Furthermore, volunteering is equally suitable for older people who already have a career but want to try out something new. It's a good way to "test the waters," or experiment with something new. After some volunteer experience, you may decide that it isn't the best career choice for you. At the very least, you come away with a clearer sense of what you want to do for a career.

Interning and volunteering are valuable, worthwhile methods for achieving one's career goals. Both allow young people to explore career positions and gain experience prior to landing an official job. In this book, we will discuss how both experiences can help people who love nature to find the perfect job for them.

# Chapter One

# WHY VOLUNTEER OR INTERN?

The thought of starting a career can be stressful. Even when you think you know what you want to do, it can be difficult to figure out how to do it. College is an obvious choice. However, many college students discover within their first few years that they may have made the wrong choice. How can young people avoid this pitfall? Volunteering and interning prior to college are two excellent ways to try out a career. However, you might be wondering how interning and volunteering are different. Which is the best option for you? Is one better than the other?

## The Benefits of Volunteering and Interning

The job market has become more competitive in recent years. More young people are vying for a smaller pool of jobs, which means that any experience in your chosen field gives you a head start. Early experience gives you a greater chance of getting interviews and landing jobs. It may also lead to a higher starting salary.

A veterinary technician intern *(left)* works with vets from the Buttonwood Zoo in New Bedford, Massachusetts, to care for a loon that had been affected by an oil spill.

Volunteer positions and internships give young people the chance to "test drive" a career. They give you valuable on-the-job experience. Internships are usually counted as credit toward graduation. For those who decide not to attend college, a volunteer experience can help make a smooth transition into a real job.

## Introduction to the Field

Many recent college graduates discover they aren't fully prepared to enter the workforce, despite the many years of hard work they devoted to their studies. As valuable as college is for the young professional, it's sometimes no match for real-world experience.

Interning and volunteering give young people a unique opportunity to gain experience and develop on-the-job skills. These avenues offer a hands-on introduction to a particular field. For example, a volunteer position with an animal shelter is a great way to begin learning how to care for animals. An internship with a botanical garden is an excellent way to make a smooth transition into the fields of botany and horticulture. Volunteering and interning give you a taste of what it's like to work in a specific field and allow you to decide if it's really what you want to do.

Forming professional relationships with others who share your interests is a great way to help prepare for a long, successful career.

## Build a Network

With the popularity of sites like Facebook and Twitter, you've most certainly heard of social networking. This is a group of friends or acquaintances linked by similar interests and activities. Anyone interested in establishing a professional career—regardless of the field—should build a professional network in much the same way that one builds a social network.

Professional networking is the creation and maintenance of business friendships and acquaintances. Your network is a bank of contacts who have the potential to help further your career in some way. Contacts in your professional network could be anyone you meet on the job or in school, or even people in your family. A contact might write you a letter of recommendation or help you write a great résumé. Or a contact might just help you get a better

# Let's Network!

Anyone you know—including parents, siblings, friends, teachers, coaches, and neighbors—could be in your network. However, the first real step in networking is meeting new people. Be a good listener and ask questions. Present yourself and your skills without being too pushy. Stay in touch with your contacts as much as you can. Call them up just to say hi.

Write notes about each contact, especially those you've spoken to only a few times. Are they married? Do they have kids? What are their hobbies and interests? These notes will come in handy. If you don't talk to a contact for several months or longer, you will still be able to connect to them on a personal level.

There will come a time when you'll ask a network contact for a favor. That's the main idea of networking and the reason why it's so important to maintain friendships. Keep in mind that networking is a give-and-take process; there will come a time when a contact will ask you for a favor. Always do your best to help others, and they will be more likely to return the favor when you need it.

It's important to build a network before you need it. In time, your network will grow. Strengthen your network by introducing contacts to each other. If you can help contacts strengthen their network, they'll be happy to repay the favor in the future.

parking spot at work. In short, maintaining a professional network is a great way to increase your chances of success.

## Discover a Mentor

A mentor is someone with considerable experience in and knowledge of a given topic. He or she is available to teach you the ropes and lend support. A mentor is often older than you, but that's not always the case. Not everyone is fortunate enough to find a mentor. However, if you can recognize mentors when they present themselves, you'll be able to benefit from the relationship and make the most of it.

Finding mentors is one of the greatest benefits of becoming a volunteer or intern. They can guide you as you learn. After you gain experience, you may even have the opportunity to mentor a less experienced person. Mentoring is an important way for an expert to pass knowledge on to a novice.

## Develop Your Résumé

A résumé is a written summary of your skills, experiences, and education. Along with a cover letter, the résumé is your introduction to new employers. Résumés need to be well crafted to make a good first impression.

Often, recent high school graduates don't recognize the importance of a strong résumé. Even for recent college graduates, it can be difficult to create a résumé that will catch an employer's attention. College experience looks great on a résumé, but that alone won't guarantee an interview. Most employers are looking for real-world experience.

Many recent graduates list jobs that may not interest employers. Jobs such as waitress or cashier may show that you've been a hard worker, but they won't necessarily interest some employers.

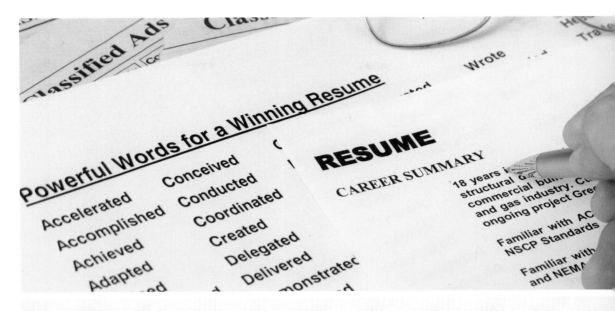

Writing your first résumé can seem difficult. However, there are numerous resources for first-time résumé writers online and at the library.

For example, the owner of a horse ranch or a state park administrator may not care if you were a waitress. Those types of employers will be looking for very specific skills. The owner of a horse ranch is more likely to interview someone who has volunteered on a farm or at a stable. A state park administrator will probably interview someone who has interned with a park ranger before they interview someone who worked as a cashier. In short, interning and volunteering are résumé gold.

## Land a Job

In most cases, a college education increases your chances of finding a job and starting a career. However, it's not the only career path available to you. Many people are able to establish successful careers without a college education. Depending on your chosen career, it's quite possible to land a dream job based

The Live Power Community Farm in Covelo, California, provides food for nearly two hundred homes in the San Francisco Bay area. It provides a valuable nine-month learning experience for young volunteers every year.

on previous jobs and previous volunteer experiences.

Many internships are part of the coursework of a college major. They are usually unpaid positions that result in college credit. However, many internships are designed to train college students for a future position with the same organization. You can even find internships before college, which can help you move into a college program or get a good job without a college education.

## Benefits for the Employer

Most employers prefer to hire workers with prior experience. Experienced individuals require less training. Employers know applicants with prior experience have learned the ropes and can handle basic skills. They've already proved themselves to be competent, knowledgeable workers.

Employers don't create internships just to be kind to students. They're hoping to evaluate and hire qualified employees. Some employers offer volunteer opportunities in the hope of preparing someone for a full-time position. Most employers will gladly interview applicants who have done everything they can to gain knowledge in their chosen professional field. This alone is a good reason to volunteer or intern.

# Chapter Two

# WHERE TO START

For many teens, starting down the road to a professional career can seem discouraging, even impossible. Even when you know what you want to do, it can be difficult to know where to start. It's not always possible to land a job in your chosen career upon graduating. However, there are numerous volunteer positions and internships out there for those who know how to find them.

There's more than one way to find an internship or volunteer opportunity. Once you know where to look, you might be surprised to discover the abundance of chances out there for young people to gain valuable experience.

## Friends and Family

The most logical step when seeking career guidance is to lean on the people closest to you. Always ask yourself who you know and what they can do for you. Often, the best source of help in career planning are your own parents. Your parents know you best and understand your

Parents and other relatives are often our most valuable mentors. They are usually very happy to give you advice and help you start preparing for a career.

hopes and dreams. Discuss with them what you would like to do, and ask for guidance.

An often overlooked source of help is friends and their parents. Perhaps your friend's mother is a guide for a local natural history museum. Or maybe a friend's father owns a nursery. It's helpful to ask these parents if they know of any volunteer opportunities or internships. Always look for chances like these to build your professional network and improve your chances of success.

## School Sources

Teachers and coaches are often a student's greatest source of encouragement and guidance. Science teachers may be aware

School counselors are professional mentors. They can help you make the best decisions for your future career path.

of local volunteer opportunities working with museums, parks, zoos, or colleges. Or they may be able to recommend a nature-based internship. English teachers may help write a letter of application to a college or internship program. In general, many teachers, regardless of their area of expertise, are very interested in helping students get a foot in the door.

Guidance counselors and advisors are trained to aid students in succeeding in high school but also in making the transition into college or a career. They help students with the college application process, coursework selection, financial aid and scholarships, and job searches. Counselors and advisors can help you find internships in your chosen career field. They will also likely have information about volunteering.

## Libraries

School and public libraries have career resources for young people. These resources are often grouped together in a common area. Librarians are always there to help you find what you need.

In addition to books on numerous topics—from careers in nature to how to write a résumé—your library will probably also have other helpful resources. Many have a community bulletin board listing jobs and volunteer opportunities, perhaps things such as local garden clubs, animals shelters, and hiking clubs. It's always helpful to scan the listings for local opportunities. You

# LEAF

Leaders in Environmental Action for the Future (LEAF) is a paid internship program for students enrolled in environmental high schools across the nation. It's sponsored by the Nature Conservancy, an organization that works to protect habitats around the world.

The environmental schools involved in the LEAF program were created to give children in urban centers the chance to learn about nature and conservation issues while teaching them career skills. They share resources and program standards to help students gain a solid understanding of the issues facing world habitats.

Students enrolled in the LEAF program get a unique opportunity to become conservation interns. They're placed in small groups led by mentors and spend four weeks in one of the many nature preserves across the United States. Interns gain experience by participating in numerous tasks, including land management, scientific research, and controlling invasive species, to name just a few. They also get the chance to enjoy nature by participating in outdoor activities, such as hiking, camping, and kayaking.

LEAF internships are designed to educate a new generation of conservation experts and give them the skills and knowledge necessary to address Earth's most urgent environmental problems. While gaining an appreciation for the natural world, interns also learn valuable career skills.

can also check out the classified sections of local and out-of-town newspapers for more volunteer opportunities.

## Online

Just about any information you need can be found online. It can be difficult to find reliable sources on the Internet, but there's no question that career advice is plentiful in cyberspace when you know where to look.

Toward the end of this book, readers will find a thorough list of dependable Web sites and organizations. These resources are geared specifically to people who love nature and are interested in volunteering and interning.

## Job Fairs

Attending job and career fairs, both in your school and your community, is an excellent way to discover information about available

National Career Fairs is an excellent resource for finding career fairs all over the country. Its Web site also offers expert advice on how to make the best of your career fair experiences.

career opportunities. They allow you to explore the local job market and see what it has to offer. Many job fairs, particularly those associated with colleges, also feature internship searches. Some organizations will even hold job fairs in the hope of finding volunteers. Check with your guidance counselor and teachers for information on job fairs in your area.

## "Door to Door"

Take the time to visit local businesses and organizations that interest you and enquire about volunteer positions. During your visit, get to know the place and even the employees working there. This allows you to establish contacts and begin pursuing opportunities. It also displays a true sense of ambition—a trait that potential employers are sure to appreciate.

For the individual who loves nature, there are numerous locations you might consider visiting. They include local parks, botanical gardens, zoos, aquariums, museums, and camps. The more you visit these places, the more you will understand how they operate. You will also get to know the staff, increasing your chances of meeting with success.

## Ready for Action

Finding a volunteer position or internship is just one of the first steps to establishing a career. Each encounter brings you one step closer to realizing your career goals. Each volunteer position or internship helps build an effective résumé and strengthens your professional network.

The following chapters will offer you concrete ideas about the types of nature-related jobs that may require interns and volunteers. Each focuses on a specific area of employment for people who love nature.

## Chapter Three

# EMBRACING THE GREAT OUTDOORS

D o you like hiking and camping? Do you enjoy visiting state parks and wildlife areas? If you do, you'll be encouraged to hear that there are plenty of internships and volunteer positions out there for people who enjoy the great outdoors.

Many of the locations and job sources discussed in this chapter are perfect for people who like to lose themselves in nature. They may thrive on the peace and quiet of the natural world itself. For nature-related positions that include monitoring, teaching, and guiding people in nature, turn to the next chapter.

## State and National Parks

In 1916, President Woodrow Wilson approved legislation creating the National Park Service. This federal agency was established to manage and maintain national parks and monuments. They were entrusted to preserve scenery, historic objects, and wildlife for future generations to enjoy. The National Park Service has more than twenty thousand

# Volunteers-In-Parks

The National Park Service maintains a very special program that invites volunteers to help protect and preserve America's natural treasures. The Volunteers-In-Parks (VIP) program gives people from all over the world the chance to volunteer their services at a national park. The VIP program has opportunities for volunteer rangers, secretaries, tour guides, researchers, computer specialists, and much more.

The many state and national parks across the United States are in constant need of young volunteers and interns. Whatever your area of interest or skill set, the National Park Service's VIP program is sure to have something for you.

employees, including rangers, biologists, electricians, landscapers, historians, and carpenters, to name just a few.

The positions available in state and national parks often depend on their location and habitat. Jobs at Yosemite National Park in the mountains of California will differ from those in Acadia National Park on an island off the coast of Maine. Arizona's Grand Canyon National Park will have different positions than Florida's Everglades National Park.

On the other hand, every park has similar opportunities for interns and volunteers. There are numerous areas of work and a multitude of skills you can learn at a state or national park. They help train geologists, biologists, archaeologists, landscapers, soil specialists, and many others. No matter where you live, there's sure to be a state park nearby where you can volunteer or find an internship. These locations feature some of the United States' most amazing natural formations, from New York's Niagara Falls

State Park to Nevada's Lake Tahoe State Park. State parks are sure to offer you the same opportunities offered by national parks. You will read even more about state and national parks in the following chapters.

## Wilderness Areas

Despite the country's widespread industrial presence, North America is a treasure trove of natural resources. The U.S. Forest Service protects and maintains 193 million acres (780,000 square kilometers) of forests and grassland. Much like state and national parks, this agency is in constant need of capable volunteers.

National forests aren't the only wilderness areas that may offer internships and volunteer opportunities. You may also find prospects at local camps, campgrounds, and nature preserves. As

A park ranger in Clarks Hill, Georgia, monitors fish passing through Lake Thurmond Dam. Rangers can teach a lot to volunteers and interns.

mentioned in chapter 2, you can find these types of opportunities by visiting local organizations, reading your local newspaper, and by researching local wilderness areas online.

# Nature and Hiking Trails

Nature trails are very popular all across the nation. You can find them in rural areas as well as the outskirts of our nation's cities. Many urban centers have even begun constructing new nature trails to give citizens a place to exercise and enjoy nature. These trails are in need of volunteers to keep them neat and litter-free. Young people may even get the chance to help create them.

The American Hiking Society has established several nationally recognized volunteer programs to help create and maintain hiking trails. Their Volunteer Vacations program gives people the chance to help build and clear hiking trails. Volunteers spend several days camping in the wilderness, working to improve public land. These outings are led by experienced crew leaders.

# Career Paths

The following sections provide information about the kinds of positions you will find in parks, wilderness areas, and trails. In addition to the positions mentioned here, parks, campgrounds, and wilderness areas are in need of experienced maintenance workers, groundskeepers, landscapers, and firefighters.

## Ranger

Rangers are the police officers of national parks and forests in the United States and Canada. Often known as forest technicians, rangers have numerous duties. They're often called on to put out forest fires. They build and maintain park and forest facilities. They

Rangers may be called upon to be spokespeople for the park or forest for which they work. They may be asked to address the public during press conferences.

help create, clear, and maintain trails. Some rangers help plant trees, remove dead trees, and even aid in the sale of timber.

Many rangers are uniformed law enforcement officers who protect the nation's natural resources. They are entrusted with monitoring natural sites and enforcing laws. These are just a few of the tasks rangers may be expected to handle on a day-to-day basis. One thing that ties all these duties together is the chance to work in the great outdoors. The U.S. Forest Service offers numerous internships and volunteer opportunities for people interested in becoming rangers in America.

## Backcountry Caretaker

Backcountry caretakers tend to trails and facilities in out-of-the way places. They clear paths and repair bridges. They maintain visitor buildings, shelters, campsites, outhouses, and other services. In short, backcountry caretakers make sure the services and facilities offered by a park are in perfect working order. Some are asked to help educate campers and hikers about local sites and dangers. Others aid in search-and-rescue operations.

Some backcountry caretakers are on duty for a week or more and then have four or five days of rest. Their headquarters while on duty might be a tent in the wilderness. Volunteer positions are common in the summer when professional caretakers need a break or when caretaker training is easiest.

## Trail Coordinator

In addition to nature trails, there are trails for hiking, running, biking, horseback riding, and snowmobiling. Experienced trail coordinators are needed to design and build these trails. They must take into account numerous kinds of terrain—from mountainous to wooded. Trail coordinators make sure trails are properly mapped

An intern uses a machine to compact the soil of a new trail in Cuyamaca Rancho State Park near San Diego, California.

and marked for the public. Volunteer positions and internships are available for trail coordinators.

## Geologist

A geologist studies the materials that make up Earth and the forces that shape it. They're experts in minerals, mining, erosion, and the study of natural disasters. Geologists are needed for a multitude of jobs, most of which entail working in the great outdoors. Some monitor volcanic and earthquake activity. Others investigate soil contamination. Still others work side-by-side with archaeologists at fossil sites.

The U.S. Geological Survey, National Park Service, and Forest Service all have positions for geologists. They also offer numerous geology internships and volunteer positions across the nation.

# Chapter Four

# TEACHING OTHERS ABOUT NATURE

Instead of losing themselves in nature, some people get great satisfaction from teaching others what they know about the natural world. There are many great career opportunities for these people. Several of the work environments mentioned in this chapter are covered in other chapters as well. However, the main emphasis here will be careers where working with the public is the key to success.

## Youth Camps

Every summer, children all across the nation attend camps in natural settings. Summer camps feature camping, hiking, swimming, athletics, and many learning experiences. Youth camps are an excellent way for teen volunteers to gain experience monitoring and teaching others while enjoying the natural world.

There are numerous kinds of youth camps. Some feature fishing, rock climbing, rafting, archery, or even wilderness survival. Some camps focus on specific sports.

Young children learn about animals from interns at farms such as Shelburne Farms Adventure Camp in Vermont.

Others focus on academic endeavors. This range of theme-based youth camps allows individual volunteers plenty of variety when choosing a camp to apply to.

## Nature Centers

Tifft Nature Preserve, just south of Buffalo, New York, features marshes, woodlands, and ponds teeming with wildlife. It has nature trails, cross-country skiing, fishing, and guided tours. It's a popular destination for school field trips as well as family day trips.

Many nature centers like Tifft are small and often overlooked by people who love nature. However, most nature centers need volunteers to help educate and work with the public just as national parks do. Some even offer internships. Are there any nature centers in your local area? You might want to visit and inquire about internships and volunteer positions.

## Museums

Natural history museums—such as the Smithsonian National Museum of Natural History in Washington, D.C.—focus on the scientific research of plants, animals, earth sciences, and paleontology. They feature plant and animal specimens, diverse habitats, fossils, rocks, minerals, meteorites, and

Interns and volunteers can assist professionals in their fields, such as Dr. Richard Benson at the Smithsonian Natural History Museum in Washington, D.C.

even human artifacts. Natural history museums employ hundreds of scientists, tour guides, and naturalists. Science museums, which you can find in most major cities, offer many of the same opportunities.

## Zoos and Aquariums

Zoos and aquariums all over the country are in constant need of interns and volunteers. Funding can be difficult for these organizations to attain, which means they appreciate any help they can get from the public and from students.

Zoos and aquariums have plenty of positions for people who want to work with the public. These include tour guides, school education programs, main desk help, and telephone operators. If

The American Museum of Natural History in New York City allows college students to volunteer to teach their own educational programs. Here, one teaches young students about blowfish during his "Attack and Defense" tour.

you've always liked going to the zoo, it might be the perfect fit for you. Check with your local zoo or aquarium to find out what it has to offer.

# State and National Parks

Volunteers and interns have many opportunities to interact with the public in state and national parks. The National Park Service in America and Parks Canada are in constant need of guides to teach visitors about a park's wildlife, history, landmarks, and sights. They also need campground hosts, counselors, cooks, and secretaries, to name just a few of the positions that are available for volunteers and interns.

Internship and volunteer positions can eventually lead to careers, such as a tour guide in Everglades National Park in Florida.

# Career Paths

Parks, camps, nature centers, museums, and zoos need responsible people to serve as clerical workers, such as secretaries, phone operators, and proofreaders. While data entry and document filing may not seem that exciting, these types of positions give a young person the perfect chance to get their foot in the door. This can lead to a more stimulating job. The following sections cover a few of the more interesting positions serving the public.

## Tour Guide

Working as a tour guide for a nature-based destination or museum can be both enjoyable and rewarding. Tour guides are educated in the subject matter and trained to teach visitors everything there is to know. They're trained to answer just about any question that may come up. They must also be able to relate interesting stories with little notice in case of delays or closed exhibits. There are numerous kinds of tours that you may find yourself leading. For example, Everglades National Park in Florida offers tours for kayaking and

# The Smithsonian
# National Museum of Natural History

The Smithsonian Institution is the world's largest museum and research complex. It's comprised of nineteen museums and galleries, the National Zoological Park, and nine research facilities. The Smithsonian also has 168 affiliate museums across the country. These educational centers feature artistic, historic, and natural treasures.

The National Museum of Natural History (NMNH) in Washington, D.C., has exhibits, collections, and educational programs to teach the public about the natural world. In addition to the main museum, the NMNH has a collection storage facility in Suitland, Maryland. The museum also has a marine science research center in Fort Pierce, Florida, and field stations in distant locations such as Belize and Kenya.

The NMNH offers numerous internships. It even has some for high school students. Some interns work directly with the public. Others aid in scientific research or museum administration. Also, volunteers are needed to lead tours, help with school programs, man information desks, and answer telephones.

canoeing, photography, fishing, bird watching, boating, and several others.

Tour guides are often referred to as "interpreters." It's an interpreter's duty to help visitors care about and care for the natural resources of the park. They create audience interest in the park, its habitats, and its wildlife. Interpreters "translate" the objects and resources into a language that visitors can understand and care about. This is a special skill, one that can be learned by interning or volunteering as a tour guide.

## Teacher

Teachers who love nature have many opportunities. Museums and parks need teachers to lead educational programs. Some have regular classes for both high school and college students. Many forest rangers are called upon to educate visitors on safety and park rules. Geologists may teach visitors about local land-forms. Some teachers become tour guides. Summer camps need people, such as camp counselors, to help teach children. Some teachers who love nature even help train interns and volunteers who love nature.

## Naturalist

A naturalist is an expert in natural history. Naturalists are like teachers in many ways. They often educate park and museum visitors on important topics. However, their expertise in the field of natural history gives them different options and opportunities.

A naturalist might be asked to plan tours for school groups and then act as a tour guide. Naturalists sometimes give lectures about park facilities and resources. They may be selected to help design and construct scientific or historic museum displays. Naturalists may also be required to organize and manage staff meetings and activities. Furthermore, they are often chosen to guide and educate interns and volunteers.

# Chapter Five

# CARING FOR ANIMALS

There are many internships and volunteer opportunities out there for animal lovers. Zoos and aquariums are obvious choices, and most of them have numerous opportunities for young people. Both need volunteers to help keep facilities clean, including animal cages. Some volunteers get to work directly with animals with the guidance of experienced professionals. Internships at zoos and aquariums may include animal observation and research, animal care, and veterinary medicine.

Zoos and aquariums aren't the only places where young people can find volunteer opportunities and internships. This chapter addresses other career choices for people who love animals.

## Animal Shelters

The North Shore Animal League (NSAL) is an animal shelter in Port Washington, New York. It's the largest "no-kill" rescue and adoption center in the world, which means it doesn't euthanize, or put to sleep, unwanted animals. This organization has rescued more than one million dogs and cats since it started in

Animal lovers can find opportunities to help at places such as the Sacramento County Animal Shelter and Adoption Center.

1941. There's no way the NSAL could rescue so many animals without the dedicated help of numerous volunteers. The NSAL needs volunteers to help train, feed, walk, and play with animals. It also needs volunteers to help educate the public on how to take care of animals.

The NSAL might be one of the most well-known animal shelters, but it's just one of millions all around the United States in need of dedicated volunteers. Do some research to find an animal shelter near you. Chances are it has opportunities for volunteers. The basic skills you learn volunteering at an animal shelter are sure to help as you plan a career helping animals.

# Animal Sanctuaries

Animal sanctuaries are havens for abused, abandoned, neglected, and endangered animals. Instead of trying to find a home for animals, as shelters do, animal sanctuaries provide a natural habitat for wild animals and allow them to live their lives in peace. The animals in a sanctuary are protected, cared for, and given the chance to behave as they would in their natural habitat. Most sanctuaries are for wild animals, although there are some for domesticated animals, such as dogs and horses.

Animal sanctuaries are expensive to set up and maintain. Many are run by volunteers who are animal lovers. Sanctuaries need dedicated volunteers to clean facilities, build and repair fences, set up habitats, and care for animals. They also need people to do clerical work and answer phones. It's hard work, but it's fulfilling. It also looks good on a résumé.

# Ranches

Ranching is the practice of raising livestock for food, cloth, and labor. Ranches are large areas of land where livestock, such as cattle and

# The Wildlife Science Center

The Wildlife Science Center (WSC) in Minnesota was established in 1976 as a federally funded research facility. It was founded to observe and record the behavior of a captive population of gray wolves. Federal funding for the project ended in 1991. Since then, the WSC has expanded by focusing on educational efforts. In addition to gray wolves, the WSC now has red wolves, bears, raptors, bobcats, and foxes. It hosts numerous educational programs for the public, including school tours and overnight camping trips.

More than twenty-five thousand people visit the WSC every year. It hosts many activities, from dog sledding to weekend tours. Without the federal funding it once received, the WSC's success relies on the dedication of hardworking volunteers.

The WSC volunteers are involved in a wide range of duties. They learn valuable on-the-job skills from trained wildlife professionals. In addition to participating in numerous research projects, volunteers can work with field professionals, animal control officers, veterinarians, zoo professionals, and wildlife managers. Volunteering for the WSC is an excellent opportunity for young people to learn about the care and protection of endangered species and the conservation of natural resources.

sheep, can graze. Some ranches raise horses, and many use horses to help tend other animals and travel long distances. Jobs on a ranch vary. Some workers tend to the animals. Others build and repair facilities. Some ranches give guided tours, and others feature hands-on activities for the public.

Many ranches offer internships. Students can learn how to care for and ride horses. They may be trained in veterinary medicine. Others shear sheep, tend to young animals, or construct fences. Some ranches—especially those that feature tours, exhibitions, and educational experiences—offer volunteer positions.

# Career Paths

Zoos and aquariums around the world need zookeepers to care for, train, and feed animals. Being a professional zookeeper can be both fulfilling and challenging. Zookeepers must be able to care for a diverse range of animals—from dangerous grizzly bears to friendly penguins. Other duties for zoo workers may include building, maintaining, and cleaning facilities, supplying medical treatment for animals, and conducting tours. Working for a zoo or aquarium might be the perfect opportunity for people who love animals. However, there are other career paths for animal lovers.

## Animal Caretaker

"Animal caretaker" is a loose term that describes anyone who tends to the well-being of animals. Most animal care positions involve feeding animals. They may also include washing the animals and the places where they live and sleep. Animal hospitals need people to help care

A volunteer for World Wide Opportunities on Organic Farms helps care for two baby pigs who lost their mother.

for patients and keep facilities clean. Many research and medical facilities need workers to care for animals used in testing. There is a caretaker job for just about any type of animal you can think of.

## Animal Rescue

Many animal shelters and rescue centers depend on volunteers. However, there are plenty of paid positions with these types of organizations, too. Local governments across the country need trained animal control specialists to help rescue animals or capture dangerous animals. Wildlife rehabilitators take in wild animals that have been injured or orphaned and nurse them back to health so that they can be returned to the wild. Shelters and rescue centers need managers, trainers, vets, and custodians.

## Veterinarians and Vet Techs

Becoming a veterinarian is hard work. It takes up to eight years of college, as well as one or more internships. Similar to regular doctors, vets may specialize in a specific area of veterinary medicine. While it isn't the best option for everyone, veterinary medicine can be a fulfilling and even lucrative career choice. You may work for an animal hospital, shelter, farm, zoo, or research facility. Or you might open your own practice.

If you're interested in veterinary medicine but don't want to attend eight years of college, you might be interested in becoming a veterinary technician, or vet tech. Vet techs are animal nurses. They help veterinarians. Vet techs require an associate's degree.

## Zoologist

Zoologists are scientists who study animals. Many zoologists monitor wildlife and the effects of human activities on them and their

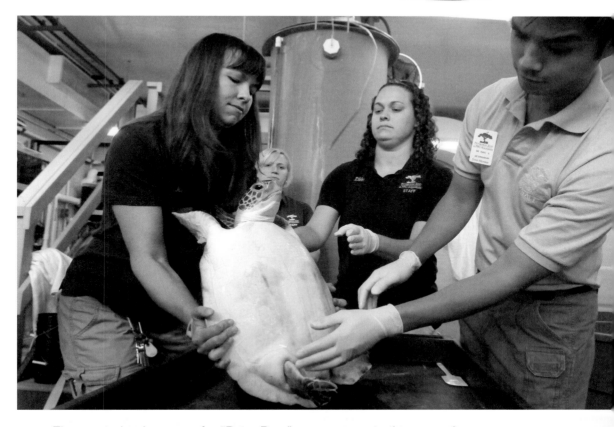

Three veterinarians care for "Peter Pan," a young sea turtle rescued from Sanibel Island, Florida.

habitats. Much like veterinarians, zoologists require many years of college. However, they can land high-paying jobs at a variety of organizations. They might work for a zoo, but they can also work for government agencies, private businesses, and large industrial companies.

Zoologists sometimes specialize in one area, such as ecology, conservation, or education. A marine biologist is a special kind of zoologist that studies animals that live in the oceans or other bodies of water. Marine biologists may work in numerous different organizations, such as fisheries, state agencies, universities, and research facilities, to name just a few.

# Chapter Six

# CARING FOR PLANTS

**N**ational and state parks need people to monitor the health of local plants. Some are trained to seek out and remove invasive plant species. Some plan and guide tours that teach visitors about park flora. Landscapers plant and tend floral displays and maintain lawns and gardens on park property.

While state and national parks may have numerous opportunities for people who love plants, they're not the only places where people with a green thumb may find a volunteer position or internship.

## Farms

Farming has long been, and still is, an essential industry for any society. Millions of farms across the United States grow the fruits and vegetables necessary to feed our nation. Others grow flowers or cotton for textiles. Many farms offer internships. Applicants often need little or no previous experience. The work can be tiring and the hours can be long, but it is also a great learning experience

An intern at the Peabody Ranch in Del Rio, California, helps harvest vegetables that will be sold to families and restaurants in the area.

Farms need interns to prepare land for seeding, planting, watering, cultivating, and transporting plants. These are just a few of the duties you might experience. Some farms produce maple syrup. Some operate markets and food stands. Others grow highly specialized crops, such as Christmas trees. Regardless of the type of farm, interns will participate in many agricultural processes from beginning to end.

## Nurseries

Nurseries specialize in the propagation and care of young plants. Most grow ornamental crops, or plants used to beautify homes and businesses. Nurseries range from small local greenhouses

# Volunteer Gardeners

Volunteers who love nature have numerous opportunities to practice their craft and learn from others. Most towns and cities have programs for amateur gardeners who want to help beautify their neighborhoods. Many schools have also developed volunteer gardening programs to teach students how to grow their own flowers and vegetables. You can even organize your own gardening program to help your community. Joining or running a gardening group is a good way to learn about horticulture and landscaping.

Urban gardening groups have gained momentum in recent years. Finding and using the limited space and natural resources in an urban environment presents unique challenges for amateur gardeners. In fact, gardening in a city can teach you many new techniques. Some urban gardeners pride themselves on creating idyllic getaways in the middle of a city. Others become masters at growing productive vegetable gardens. Urban gardening volunteers help transform vacant lots into beautiful parks and gardens. They can also help feed local communities while teaching residents about the importance of protecting and cultivating our natural resources.

Habitat gardening is a way to provide an ideal habitat for local wildlife. A habitat garden is filled with plants that make native animals feel at home. They can be designed to attract butterflies, hummingbirds, and more. While many people create habitat gardens in their backyards, creating and maintaining a habitat garden is a worthwhile communal effort. By forming or participating in a neighborhood habitat garden, volunteers help improve their community while learning about conservation issues.

that sell to the public, to large corporations that sell to other businesses.

Much like farms, many nurseries offer internships. Interns get the chance to learn about seeds, planting techniques, plant health, soils, and propagation. The skills interns learn will help them pursue a career in plant production and sales. Many will go on to open their own greenhouses and nurseries.

# Botanical Gardens

Founded in the Bronx, New York, in 1891, the New York Botanical Garden contains fifty gardens and plant collections. It has more than one million plants. The centerpiece of the botanical garden is the Enid A. Haupt Conservatory, a giant glass-walled building housing plants from eleven different habitats. This historic natural location offers many internships to college students each season.

Botanical gardens are fascinating and often relaxing places to visit. Many specialize in a certain kind of plant; an arboretum is a botanical garden that specializes in trees. Many botanical gardens feature classes, guided tours, and scientific research projects as well. There is much to learn about plant life and plant care by interning or volunteering at a botanical garden.

# Career Paths

There are many career opportunities for people who love plants. Some may decide to open a flower shop, while others may operate an apple orchard. Even farming offers a wide range of prospects based on the individual's interests. Traditional farming is still a popular choice, but organic farming is growing bigger as more people become interested in protecting the environment. The following sections address a few of the more popular career paths for people who love nature.

Interns and volunteers can learn from knowledgeable and experienced professionals, such as this botanist with the U.S. Forest Service and Department of Agriculture.

## Botanist

Botanists are scientists who study plants. They also study how environmental factors affect plant growth. Botanists often special-ize in one type of plant or in a single method for studying plants.

Becoming a botanist requires a college degree, but there are many career paths a botanist may pursue. Many become teach-ers. Others research how plants can be used in the manufacture of foods, fibers, drugs, and other products. Botanists can find work in many of the organizations already mentioned in this book. National parks and forests need botanists to help man-age plant resources, investigate invasive species, and study the effects of human activities on native plant species. Many botanists work for botanical gardens, museums, and nature centers, too.

## Forester

Foresters cultivate and maintain our nation's for-ests. They protect existing forests and help reforest areas where trees have been removed due to over-logging. Foresters plant new seedlings and remove old and dying trees. They battle insects and diseases that attack trees. They also identify the causes of soil erosion in our forests and work to reverse its negative

Hands-on experience is a great benefit of interning or volunteering in a specific field, such as forestry.

effects. In some cases, a forester's job might be very similar to a ranger's.

Many foresters work for national parks or forestry services. They're also employed by nurseries, orchards, and specialty farms. Foresters must have a college degree. Internships are common and can be found in forests all over the nation.

## Horticulturist and Landscaper

Horticulture is the cultivation of fruits, vegetables, flowers, or ornamental plants. Horticulturists are experts at cultivating plants. Job

Horticulturists, such as this one at Mellon Park Walled Garden in Pennsylvania, can offer great insight to student volunteers who are considering a career in the field.

prospects for horticulturists are very good. Similarly, landscapers are gardeners who specialize in maintaining large gardens and lawns. You may be familiar with landscapers in your local neighborhood. However, many landscapers do more than mow lawns and trim bushes. Some plan, create, and maintain the facilities for parks, camps, and wildlife preserves. They're often called landscape architects.

State and national parks hire horticulturists and landscapers to maintain trails, flower beds, lawns, and other facilities. Local governments hire them to maintain town and city properties. They also work for nurseries, botanical gardens, wilderness areas, and golf courses, to name just a few.

# Chapter Seven

# CARING FOR HABITATS

**E**arth's natural habitats face many dangers because of the activities of people. Acid rain caused by dangerous pollutants affects many kinds of plants and animals. Marine animals are threatened by overfishing, ship traffic, and pollution. Rain forests shrink every year because of logging and farming. The climate of the Arctic has changed so much in recent years that animals like the polar bear face new challenges they may not be able to overcome. Habitats, much like the plants and animals that live in them, need our help.

Despite environmental problems such as those mentioned above, many people and organizations are taking great steps to help Earth heal. State and national parks and national forests have become training grounds for our most concerned and dedicated conservationists. Park employees—including scientists, rangers, foresters, veterinarians, and naturalists—tirelessly research environmental issues, monitor native species, and protect natural habitats for future generations. Parks and forests are just a few of the many organizations and movements where volunteers and interns can learn from experienced conservation professionals.

This student is a volunteer for the Student Conservation Association. She is helping a biologist in an effort to reduce invasive lake trout in Yellowstone Lake.

# Nature Preserves

A nature preserve is a wildlife area established by a government agency or research institution. These natural areas are set up to preserve native flora, fauna, and geological features. They are designed to keep the natural habitat intact, allowing the wildlife to flourish unhindered.

Nature preserves exist in countries all over the world. Some are open to the public. Others have limited areas for public tours. Most nature preserves are used by scientists for research purposes. Rangers, geologists, biologists, and others help monitor the natural habitats of nature preserves and help educate the public on their findings. Most nature preserves have opportunities for interns and volunteers.

# Land Management

The Bureau of Land Management (BLM) is a government agency whose purpose is to sustain the health of public lands in the United States. It oversees the use and preservation of U.S. land and public resources for many purposes, including agriculture, recreation, energy development, and forest growth. The BLM employs thousands of workers, including biologists, firefighters, engineers, geologists, foresters, and rangers.

The BLM is just one of many organizations concerned with land management, or supervising the use and development of land resources in both rural and urban locations. Farms, ranches, golf courses, and construction companies must observe conservation laws while finding the optimal use for land. These businesses and organizations, including the BLM, are in need of volunteers to help professionals do their jobs. Many also offer internships for college students.

# California Conservation Corps

The California Conservation Corps (CCC) is a state agency that hires young people for up to a year to help protect and restore California's natural environment. The CCC trains new members in several areas, including conservation, disaster response, and the use of power tools. It also provides further education in specific areas depending on the member's placement within the corp.

Daily work for members may include habitat preservation, trail creation and maintenance, and producing clean energy sources. Some members aid people and animals during natural disasters, such as floods, fires, and earthquakes. CCC work is often challenging, and the conditions can be tough. However, the skills members gain far outweigh the difficulties of their jobs. In addition to promoting environmental awareness in young people, the CCC strives to encourage young people to become strong, confident workers.

The CCC offers full-time, minimum-wage positions. However, it functions similarly to an internship. Members receive special training and on-the-job experience over the course of a year. They learn skills that will allow them to establish a career in conservation, habitat restoration, and disaster response. Furthermore, the CCC is based on the philosophy of volunteerism and aiding of the local community.

# Habitat Conservation

Habitat conservation is a specific kind of land management. It is a practice that seeks to restore, protect, and conserve natural habitats for plants and wild animals. This is particularly important in areas that are home to threatened and endangered species. Habitat conservation has become increasingly important as farms, factories, urban areas, and land developers seek to claim more natural land for recreational and commercial growth.

Student Conservation Association volunteers help collect fish for study from the Little Pigeon River in the Great Smoky Mountains National Park, Tennessee.

The people who work in habitat conservation defend wildlife and natural habitats against commercial expansion. However, they also strive to reach a middle ground where human growth can be achieved without harming the natural world. For example, biologists and zoologists study habitats to determine where human growth may take place without disrupting migration patterns of wild animals. Botanists and foresters study how new industries affect the plant life in an area and suggest changes to conservation laws based on their results. Marine biologists and water specialists monitor lakes and rivers near new industries for the same reasons.

Volunteerism is very important to the habitat conservation movement. Also, many corporations and organizations have internships in habitat conservation. Young people may become involved with beach recovery, invasive species studies, seed and plant collection, and reforestation programs, to name just a few opportunities.

## Career Paths

Many of the careers addressed previously in this book play a role in the restoration and preservation of natural habitats. Following are a few career paths that play a particularly important role in these areas. Except for the final section, the information that follows focuses on positions that require a college degree. However, there are many internships and volunteer positions available to help you prepare for the future.

## Biologist

Biologists are scientists who study living things. They play many roles in habitat conservation. Research biologists closely monitor wildlife to determine the impact of environmental changes on

Near Venice, Louisiana, volunteers do the noble work of rescuing a brown pelican that is covered with oil from the *Deepwater Horizon* oil spill.

plants and animals. They design experiments to determine how future changes will affect a habitat. Wildlife biologists study the interaction of plants and animals to better understand their survival needs. Some biologists help habitats recover from natural disasters, accidents, and injuries caused by human activities. Other biologists are specialists. Marine biologists study marine life and habitats. Ornithologists study birds. Botanists are plant biologists.

## Hydrologist

The study of the movement, circulation, and quality of Earth's water is called hydrology. Hydrologists work in numerous environments—

from lakes and oceans to glaciers. They also work hand-in-hand with geologists and meteorologists.

Hydrologists are needed in the areas of land management and habitat conservation. They collect and test water samples to determine how local industries affect water supplies. They advise others on the safest and most efficient ways to use water resources. They also help protect watersheds, predict floods, study water pollution, and help forecast droughts. Others work in water treatment.

## Soil Scientist

Soil scientists study the physical, chemical, and biological elements of soil. They use this knowledge in a variety of circumstances. They often aid in land reclamation and restoration efforts. Some help farmers plan crop rotation, manage the use of fertilizers and chemicals, and conserve water. Some soil scientists work with land developers to determine the best use of natural resources. Others conduct research on soil erosion and the preservation of natural habitats. Most soil scientist positions involve fieldwork in numerous types of locations, from grasslands to urban parks.

## Field Crew Leader and Coordinator

Are you a natural-born leader? Land management and habitat conservation efforts require field crew leaders and coordinators, many of whom are volunteers and interns themselves. The Student Conservation Association, for example, needs field leaders to organize, train, and guide crews of high school volunteers. Leaders must also be able to mentor, motivate, and inspire them as well.

Field coordinators are needed to help organize, schedule, and execute activities for field crews. These positions provide a great opportunity to gain experience with natural habitats while building valuable leadership skills for a career in administration.

# Chapter Eight

# MAKING THE MOST OF VOLUNTEERING AND INTERNING

Finding a volunteer position or internship is a great way to gain experience in your chosen field. But finding one is just the first step. In many ways, these experiences are much like a real job, and you should treat them as if they were.

## Volunteering

By its very nature, volunteering requires you to freely give your time, effort, and knowledge to benefit others. A community benefits from the efforts of volunteer gardeners. Museum visitors benefit from the help and expertise of volunteer guides. Wild animals benefit from the aid of volunteer habitat conservationists.

However, there's more to gain from volunteering than making others happy. As already mentioned, there is much for the volunteer to gain as well. As such, it's in the volunteer's best interest to treat the position like a regular job. Don't be late to projects and meetings. Respect other volunteers and organizers. Concentrate on your work to do the best job you can do. By being a responsible and hardworking volunteer, you'll earn

By now, you should realize that there's a long list of volunteer opportunities available to people who love nature. The sky's the limit!

the respect of others and help increase your chances of furthering your career.

A volunteer position is a fulfilling and valuable learning experience in many ways. Not only will you learn more about the career you hope to pursue, you'll also discover more about the professional world in general. You'll learn how to interact with others and how to succeed in a professional environment. As you progress, don't forget to work on your network as well as your résumé.

## Interning

Interning can be very similar to volunteering in many ways. Much like volunteering, interning is a chance to develop your career skills and prepare for success in the real world. Treat coworkers

# Revisiting Résumés

By now, you probably have a better idea of how interning and volunteering make you a stronger job applicant. Employers are simply more likely to give you a chance with these types of experiences. There's no better way to emphasize this point than to show what a potential employer sees when he or she looks through a stack of résumés.

Let's say that you're a landscape architect in need of an assistant. The position involves a lot of hard work, including mowing, planting, and hauling supplies. However, it's also a great learning experience and a stepping-stone to a career working in nature. Brad and Jen are both applying for the job, and they send their résumés to you. Based on the following excerpts from the résumés below, which applicant would you want to interview more? Who do you think has a greater chance of landing the job based on these résumés alone?

### Jen's résumé

**Education:** high school diploma
**Previous jobs:** horse groomer, waitress
**Other experience:** flute player, gymnastics teacher

### Brad's résumé

**Education:** high school diploma
**Previous jobs:** pizza delivery, lawn mower
**Other experience:** urban gardening participant, volunteer at a local nursery

and bosses with the respect they deserve. Never forget why you are there, which is to learn.

Compared to volunteering, however, interning is often a more formal experience. While both are valuable, interns simply have more to lose than volunteers. Internships are usually parts of school curriculums. Failing to do well can result in failing a course. Not only is that a waste of your time and money, but it's a waste of the employer's time as well. An internship is an experience designed to bridge the gap between school and a career. Not everyone is fortunate enough to participate in an internship. Those who are need to realize how valuable the experience can be in establishing a successful career. Once you recognize this, you are better prepared to succeed in your internship and beyond.

World's Largest Professional Network | LinkedIn

http://www.linkedin.com/home?rk=hb_home

Linked **in**.

**Be great at what you do.**

Get started – it's free.
Registration takes less than 2 minutes.

First Name

Last Name

Email

Password (6 or more characters)

**Join Now** By clicking "Join Now" or using LinkedIn, you agree to our User Agreement and Privacy Policy.

"I make my living through the relationships garnered utilizing LinkedIn."
Kevin L. Nichols - Principal at KLN Consulting Group

LinkedIn (http://www.linkedin.com) is a business-related social networking site. It is a great way to make connections, search for volunteer and internship openings, post your résumé, and learn more about companies you're interested in.

# Create a Positive Digital Footprint

With all of this talk about résumés and networking, it would be a mistake to overlook the power of digital media and the Internet when preparing for future careers. Many teens are very involved with online social networks, such as Facebook. This is a great way to stay in touch with friends and family, but it's also a great way to create an online portfolio for employers to view.

Interns and volunteers can eventually rise to working full-time jobs in the field that they love.

An online portfolio is a collection of digital items that may include a résumé, references, personal reflections, images, and more. It is the perfect place for volunteers to create entries reflecting on their experiences.

It's important to remember that anything you put on the Internet stays on the Internet regardless of whether you delete it. This includes positive educational experiences, as well as rumors and silly pictures of you! Employers and other professionals are using the Internet more and more to research applicants. This is one of the best reasons for creating and maintaining a positive digital footprint. Much like the physical footprint you leave behind as you walk along a beach, a digital footprint is the evidence you leave behind as you interact with people online. A thorough online portfolio is one way of creating a positive digital footprint.

## Go Get 'Em!

Earth's resources and natural habitats are in need of dedicated workers to help protect and preserve them. Young people interested in nature and wildlife are encouraged to pursue the avenues discussed in this book. There's something out there for every possible interest, whether you're into animals, plants, natural environments, or just losing yourself in nature. Follow the advice in this book to find the perfect internship or volunteer position, and you're sure to meet with success in the natural world.

# Glossary

**affiliate** A group that is closely associated with a larger group.

**applicant** Someone who has applied to something, such as a college.

**associate's degree** A degree earned after two years of college.

**catastrophic** Causing widespread damage.

**clerical** Relating to routine office work.

**conservatory** A building with glass walls and roof where plants are grown and displayed.

**domesticated** Having to do with tame animals that live near or with people.

**environmental stewardship** The concern for and the practice of preserving natural resources for future generations.

**habitat** A natural setting where a plant or animal lives.

**invasive species** Non-native species that grow rapidly and disrupt a natural habitat.

**paleontology** The study of fossilized plants and animals.

**propagation** The process of reproducing plants.

**raptor** A bird of prey.

**sanctuary** An area where wildlife is protected from predators and hunters.

**specimen** A plant or animal used as a scientific representative of others of its kind.

**watershed** An area of land that drains into a river, lake, or ocean.

**zoological** Relating to the study of animals.

# For More Information

American Hiking Society
1422 Fenwick Lane
Silver Spring, MD 20910
(800) 972-8608
Web site: http://www.americanhiking.org
This is the United States' premier organization dedicated to
creating and preserving hiking trails. Through the Volunteer
Vacations program, it helps maintain trails across the
country.

Environment America
44 Winter Street, 4th Floor
Boston, MA 02108
Web site: http://www.environmentamerica.org
This citizen-funded organization researches environmental issues
facing the world today and educates the public about them.
It offers numerous internships.

Geological Society of America
P.O. Box 9140
Boulder, CO 80301-9140
Web site: http://www.geosociety.org
Established in 1888, this organization works to promote the
growth of Earth sciences and the people who choose them
as a career path.

National Museum of Natural History
P.O. Box 37012 Smithsonian Institution
Washington, DC 20013-7012
Web site: http://www.mnh.si.edu

Internship information: http://www.nmnh.si.edu/rtp/other_opps/
internintro.html

Volunteering information: http://www.mnh.si.edu/education/
volunteering/index.html

Part of the Smithsonian Institution, this museum is home to
millions of wildlife and mineral specimens and home to
nearly two hundred natural history scientists.

National Park Service
1849 C Street NW
Washington, DC 20240
Web site: http://www.nps.gov
Volunteer opportunities: http://www.nps.gov/gettinginvolved/
volunteer/index.htm

This government agency protects national parks and preserves
their natural resources.

National Zoological Park
3001 Connecticut Avenue NW
Washington, DC 20008
Web site: http://nationalzoo.si.edu
The National Zoological Park is part of the Smithsonian
Institution and one of the oldest zoos in the United States.

Nature Conservancy
4245 North Fairfax Drive, Suite 100
Arlington, VA 22203-1606
Web site: http://www.nature.org
The Nature Conservancy is an organization dedicated to the
conservation of lands and waters on which all life depends.

Student Conservation Association
P.O. Box 550

689 River Road
Charlestown, NH 03603
Web site: http://www.thesca.org
This association trains young people to protect and restore
national parks, marine sanctuaries, cultural landmarks, and
community green spaces in all fifty states.

U.S. Forest Service
1400 Independence Avenue SW
Washington, DC 20250
Web site: http://www.fs.fed.us
An agency of the U.S. Department of Agriculture, the service
maintains and protects the nation's 155 national forests and
20 national grasslands.

USGS National Center
12201 Sunrise Valley Drive
Reston, VA 20192
Web site: http://www.usgs.gov
The U.S. Geological Survey is a scientific agency that studies the
landscape of the United States, its resources, and the
dangers that threaten them.

## Web Sites

Due to the changing nature of Internet links, Rosen Publishing
has developed an online list of Web sites related to the subject of
this book. This site is updated regularly. Please use this link to
access the list:

http://www.rosenlinks.com/FID/Natu

# For Further Reading

Currie-McGhee, Leanne. *Protecting Ecosystems*. Ann Arbor, MI: Cherry Lake Publishing, 2009.

Ferguson Staff. *Discovering Careers: Environment*. New York, NY: Ferguson, 2010.

Ferguson Staff. *Discovering Careers: Nature*. New York, NY: Ferguson, 2010.

Glazlay, Suzy. *Learning Green: Careers in Education*. New York, NY: Crabtree Publishing, 2009.

Glazlay, Suzy. *Managing Green Spaces: Careers in Wilderness and Wildlife Management*. New York, NY: Crabtree Publishing, 2009.

Glazlay, Suzy. *Re-Greening the Environment: Careers in Cleanup, Remediation, and Restoration*. New York, NY: Crabtree Publishing, 2011.

King, Zelda. *Examining Backyard Habitats*. New York, NY: PowerKids Press, 2009.

Levy, Janey. *Making Good Choices About Conservation*. New York, NY: Rosen Central, 2009.

Maczulak, Anne E. *Conservation: Protecting Our Plant Resources*. New York, NY: Facts On File, 2010.

McKay, Kim, and Jenny Bonnin. *True Green Kids: 100 Things You Can Do to Save the Planet*. Washington, DC: National Geographic, 2008.

National Geographic Staff. *National Geographic Kids National Parks Guide U.S.A.* Washington, DC: National Geographic Society, 2012.

Owen, Ruth. *Growing and Eating Green: Careers in Farming, Producing, and Marketing Food*. New York, NY: Crabtree Publishing, 2010.

Reeves, Diane Lindsey. *Career Ideas for Kids Who Like Animals and Nature*. New York, NY: Ferguson, 2007.

Rohmer, Harriet. *Heroes of the Environment: True Stories of People Who Are Helping to Protect Our Planet*. San Francisco, CA: Chronicle Books, 2009.

Rosenberg, Pam. *Global Perspectives: Watershed Conservation*. Ann Arbor, MI: Cherry Lake Publishing, 2008.

Roza, Greg. *Great Networking Skills*. New York, NY: Rosen Publishing, 2008.

Somervill, Barbara. *Marine Biologist*. Ann Arbor, MI: Cherry Lake Publishing, 2010.

# Bibliography

ASPCA. "Animal Careers." Retrieved February 9, 2012 (http://www.aspca.org/about-us/faq/animal-careers.aspx).

Buffalo Museum of Science. "About Tifft." Buffalo Museum of Science. Retrieved February 16, 2012 (http://www.sciencebuff.org/tifft-nature-preserve/about-tifft).

Bureau of Labor Statistics. "Forest and Conservation Workers." Department of Labor. Retrieved February 24, 2012 (http://www.bls.gov/oco/ocos350.htm).

Dummies.com. "Considering Why People Volunteer." Retrieved January 23, 2012 (http://www.dummies.com/how-to/content/considering-why-people-volunteer.html).

Duse, Eleanor. "How to Volunteer at a National Park." HowStuffWorks.com, June 4, 2009. Retrieved February 7, 2012 (http://money.howstuffworks.com/economics/volunteer/opportunities/volunteer-national-park.htm).

Ferguson Staff. *Careers in Focus: Landscaping & Horticulture.* New York, NY: Ferguson, 2008.

Green Careers Guide. "Trail Coordinator." Retrieved February 13, 2012 (http://www.greencareersguide.com/Trail-Coordinator-Great-Job-for-Those-Who-Love-to-Hike-and-Enjoy-the-Wilderness.html).

Greenland, Paul R., and AnneMarie L. Sheldon. *Career Opportunities in Conservation and the Environment.* New York, NY: Ferguson, 2008.

Hering, Beth Braccio. "Why Are Internships So Important?" CareerBuilder, March 1, 2010. Retrieved January 23, 2012 (http://msn.careerbuilder.com/Article/MSN-2202-College-Internships-First-Jobs-Why-Are-Internships-So-Important).

Kanter, Beth. "Pumping Up Your Professional Network." SocialEdge.org, October 2008. Retrieved February 2, 2012 (http://www.socialedge.org/discussions/marketing-communication/pumping-up-your-professional-network).

Mackintosh, Barry. "The National Park Service: A Brief History." History E-Library. Retrieved February 7, 2012 (http://www.cr.nps.gov/history/hisnps/NPSHistory/npshisto.htm).

Miller, Louise. *Careers for Animal Lovers & Other Zoological Types*. New York, NY: McGraw-Hill, 2007.

Miller, Louise. *Careers for Nature Lovers & Other Outdoor Types*. New York, NY: McGraw-Hill, 2008.

Moran, Matthew. "Professional Networking Made Easy: Priming the Pump." Cisco Press, July 21, 2006. Retrieved January 30, 2012 (http://www.ciscopress.com/articles/article.asp?p=486105).

National Museum of Natural History. "About the Museum." Smithsonian Institution. Retrieved February 15, 2012 (http://www.mnh.si.edu/about.html).

National Park Service. "Alternative Guided Tours." Everglades National Park. Retrieved February 12, 2012 (http://www.nps.gov/ever/planyourvisit/alternativetours.htm).

National Park Service. "Foundations of Interpretation Curriculum Content Narrative." Interpretive Development Program, March 1, 2007. Retrieved February 11, 2012 (http://www.nps.gov/idp/interp/101/FoundationsCurriculum.pdf).

National Park Service. "Volunteer." Retrieved February 7, 2012 (http://www.nps.gov/gettinginvolved/volunteer/index.htm).

National Park Service. "Work with Us." Retrieved February 7, 2012 (http://www.nps.gov/aboutus/workwithus.htm).

National Zoological Park. "Internships and Fellowships." Smithsonian Institution. Retrieved February 13, 2012 (http://nationalzoo.si.edu/UndergradInternships/default.cfm).

The Nature Conservancy. "LEAF: Leaders in Environmental Action for the Future." Nature.org. Retrieved February 2,

2012 (http://www.nature.org/aboutus/diversity/leaf/index.htm).

Newsdesk: Newsroom of the Smithsonian Institution. "Visitor Statistics." Smithsonian Institution. Retrieved February 15, 2012 (http://newsdesk.si.edu/about/stats).

New York Botanical Garden. "Grounds and Gardens." Retrieved February 12, 2012 (http://www.nybg.org/about/grounds_and_gardens.php).

North Shore Animal League. "Who We Are." Retrieved February 14, 2012 (http://www.animalleague.org/about-us/who-we-are).

O*NET Online. "Park Naturalists." Retrieved February 13, 2012 (http://www.onetonline.org/link/summary/19-1031.03).

USDA Forest Service. "Forestry Technician." Retrieved February 13, 2012 (http://www.fs.fed.us/fsjobs/pdf/forestry_technician.pdf).

USDA Forest Service. "Geologist." Retrieved February 13, 2012 (http://www.fs.fed.us/fsjobs/pdf/geologist.pdf).

Wild Animal Sanctuary. "Volunteer." Retrieved February 18, 2012 (http://www.wildanimalsanctuary.org/waystohelp/volunteer.html).

Wille, Christopher M. *Careers in Forestry Careers*. Chicago, IL: VGM Career Books, 2004.

Williams, Nichole. "Why Volunteering Is Good for Your Career." LinkedIn.com, September 7, 2010. Retrieved January 23, 2012 (http://blog.linkedin.com/2011/09/07/profile-volunteer-field).

# Index

## About the Author

Greg Roza has written and edited educational materials for children for the past twelve years. He has a master's degree in English from the State University of New York at Fredonia. Roza has long had an interest in scientific topics and the natural world and spends much of his spare time hiking and camping. He lives in Hamburg, New York, with his wife, Abigail, and his three children, Autumn, Lincoln, and Daisy.

## Photo Credits

Cover Marvi Lacar/Getty Images; pp. 4, 8–9, 32, 33, 34–35, 42–43, 50, 51, 55, 58, 66 © AP Images; p. 10 FogStock/Thinkstock; p. 13 iStockphoto/Thinkstock; pp. 14–15 Kitra Cahana/Getty Images; p. 17 Clarissa Leahy/Cultura/Getty Images; p. 18 SW Productions/Photodisc/Getty Images; p. 20 Bloomberg/Getty Images; p. 24 © Augusta Chronicle/ZUMA Press; pp. 26–27 AFP/Getty Images; p. 28 © John Gastaldo/U-T San Diego/ZUMA Press; pp. 30–31 Melanie Stetson Freeman/The Christian Science Monitor/Getty Images; p. 39 Leilani Hu/Sacramento Bee/ZUMA Press; pp. 45, 52 © Pittsburgh Post-Gazette/ZUMA Press; p. 47 © Sacramento Bee/ZUMA Press; p. 60 Saul Loeb/AFP/Getty Images; p. 63 © The Commercial Appeal/ZUMA Press.

Designer: Michael Moy; Editor: Nicholas Croce;
Photo Researcher: Amy Feinberg